THE
# *F*INE
# *A*RT
## OF
# Technical
# Writing

# THE *Fine Art* OF Technical Writing

Key Points to Help You Think Your
Way Through Writing Scientific or
Technical Publications, Theses,
Term Papers, & Business Reports

## Carol Rosenblum Perry

Blue Heron Publishing, Inc.

# The Fine Art of Technical Writing

Library of Congress Catalog Card Number 91-72119
ISBN: 0-926085-24-X

**Publisher's Cataloging-in-Publication Data**

Perry, Carol R.
    The fine art of technical writing / Carol Rosenblum Perry.
    p. cm
    ISBN 0-926085-24-x

    1. Technical writing.    I. Title

P701                              808.066              91-72119
                                        QBI91-1306

Cover graphics and design: Marcia Barrentine
Electronic publishing and inside design: Dennis Stovall
Editorial staff: Dennis Stovall, Linny Stovall, Bill Woodall, Carol
        Hamilton

First Edition, First Printing
Printed in the United States of America on acid-free stock

Blue Heron Publishing, Inc.
24450 N.W. Hansen Road
Hillsboro, Oregon 97125
503.621.3911

In memory of my father,
Nathan Rosenblum

For my mother,
Janet Lurie Rosenblum

Dedicated to all my authors, from whom
I've learned a lot

# Table of Contents

## Conciseness — the Body Mass: Making Every Word Count   69

## Vigor — the Muscle Tone: Empowering Your Words   81

## Ending: After Writing the Last Word   101

# ACKNOWLEDGMENTS

"[Language] opens the secret gate."
— Maia Small, age 6

Gerry Spencer, a publishing veteran, encouraged me to proceed with this project when it was little more than a glimmer in my eye.

Logan Norris and Mike Amaranthus each gave me opportunities to present much of the material that would later become this book to forestry students and professionals. Preparing for those presentations forced me to clarify what I thought I knew.

Sue Borchers and Gretchen Bracher gave me crucial early feedback that helped shape the book's emphasis on making order.

John Beuter, Sue Borchers, Ed Huth, Ralph McNees, David Perry, and Sandy Ridlington took time out of jammed schedules to review later drafts. Fresh to the manuscript, they saw what I couldn't, which prompted me to rethink — and revise. Their candor was exceeded only by their diplomacy.

Carol Hamilton's keen eye scrutinized as only a well-practiced editor's can. She did to me what I've done to authors for years — and did it admirably, for which I'm grateful.

Ralph McNees has been generous above and beyond the call in accommodating the flexible work schedule that has allowed me to write as well as edit.

Jo Alexander pointed me in the direction of Dennis and Linny Stovall, whose blend of creative energy and business savvy is what most writers can only daydream of in a publisher.

David Perry, through his work on ecosystems, has taught me that language, too, is a system, subject to all system dynamics — which is the keystone to this book. His confidence in my work is one of the few certainties.

Nat Rosenblum long ago showed me that language is magical, and Jo Anne Lattin more recently showed me why.

# Prologue: Why This Book?

With the many technical writing texts and self-help books on the market, you may well wonder: why *this* book?

This little book is the logical outgrowth of my thirteen years' experience editing technical manuscripts in the natural sciences, primarily forestry — scientific articles slated for journals or other professional outlets, individual book chapters and whole books, computer documentation and user manuals. In it — distilled — are key points about technical writing that can help new or struggling authors markedly improve their writing and more polished authors refine theirs. In fact, these are the very points I've tried to convey as an editor to authors.

The book differs from others in four ways:

1  **It's not specialized.** I view technical writing broadly — as *all writing other than fiction.* So whether you're a scientist writing a research paper or grant proposal, an engineer writing technical specifications, a systems analyst writing computer documentation, a consultant writing a report for a client, a student writing a term paper or thesis, or a concerned citizen writing a letter to

your senator, this book has something to offer you because the same writing principles apply.

2  **It's selective.** The book spotlights the key points about writing. It doesn't pretend to tell you everything (you might not use it if it did) or claim to cure all writers' ills (there are no such cure-alls). It's meant to be a handy desk reference, not a dustcatcher — first to be read in one sitting and then dipped into at will.

3  **It emphasizes "making order."** Much of the book's core is devoted to helping writers who are not inherently organized about language to order their thoughts, first in their minds and then on paper.

4  **It's about more than technique.** This book is as much about the psychology of approaching technical writing and the artistry of shaping it as it is about technique. In fact, its very premise — that technical writing is a *creative* act — is meant to debunk the myth that the process of writing about "dry" subject matter is itself mechanical. It is anything but: *to write well, you must think, and thinking is never mechanical.*

Through this little book, I hope to help you recognize both the structural pitfalls that weaken technical writing and the misguided attitudes that deaden it.

# Technical Writing: the Agony & the Ecstasy

We learn language naturally — almost magically — as children listening to others speak. No one teaches us as toddlers about "grammar." Yet we in fact know a lot about parts of speech, subjects and predicates, clauses, phrases, and so on long before we hear those terms years later in a classroom. We know all this innately, unconsciously, *implicitly,* because our brains are wired to infer and internalize the deep structure of language — the grammar — through speaking and being spoken to.

Why, then, when speaking comes so relatively effortlessly, should writing — and especially technical writing — be such hard work? And it is. For nearly everyone.

**Technical writing is hard work** because we as writers must know something *explicitly* about grammar. That is, we must consciously learn something about "the rules" and "the exceptions" and, oh my, the deadly "gray areas." Then, more demanding still, we must be able to apply this knowledge of grammar with discretion — that is, case by case. You cannot write by recipe: to write, you must think.

**Technical writing is hard work** because *writers* lack some of the benefits that *speakers* have. Although writing and speaking both rely on the same words and the same grammar, they differ in critical ways. Speaking has a relatively loose structure that depends heavily on body gesture, facial expression, vocal inflection, and — except for film or videotape — the immediacy of the audience. But writing requires a far more faultless structure because it must be able to stand alone. No props, and a remote audience.

**Technical writing is hard work** because the ideas most writers are working to convey are a complexly related, nonlinear mental pattern that does not readily reduce to a "linear display" in which word follows word, sentence follows sentence, paragraph follows paragraph, and so on. In writing, content is built through nested sets of ideas, form through nested sets of grammatical constructions. And all nested sets feed back to one another. A bit esoteric? Visualize your information as a ball of string. Then imagine locating your starting point — the end of the string buried deep inside the mass of the ball — and pulling it slowly, carefully, from the mass so that the ball of string unravels *without tangling*. That's conceptually what you must surmount to write well.

**Technical writing is hard work** because it is a true creative act. The subject matter may seem dry or the manner of presentation uninspired, relative to that of poetry or fiction or screenplay. But the mental process is anything but mechanical. This is the business of the artist: to take a loosely ordered collection of ideas and transform them into something greater than the simple sum of the parts, something with shape and meaning.

# Starting:
# Before Writing
# the First Word

# Defining Your Audience

As you begin to collect your thoughts, define in your own mind the intended audience for what you're about to write. Knowing whom you're writing for determines both what you'll say and how you'll say it. But remember, although you're writing first and foremost for your intended (or primary) audience, you also probably have an unintended (secondary) audience — a more peripheral set of readers.

For example, if you're a forest ecologist writing up a research paper reporting results of your latest experiment, your intended audience may be other forest ecologists. But your unintended audience is likely to include other forest researchers such as botanists or wildlife biologists, public and private forest managers, and members of environmental or other special-interest groups. Or if you're a state law-enforcement official writing up the recommendations of a special committee appointed by the governor to select the best site for a new maximum-security prison, your intended audience may be the governor's office. But your unintended audience will almost certainly include other prison-system and law-enforcement officials, legislators, criminologists, personnel in the police departments and city or county governments of the potential sites, and the local citizenry.

Remember also that even though your intended readers are knowledgeable in your area of expertise, *they don't reside in your brain*. They haven't been with you, step by step, as your ideas have evolved; they aren't intimate with all the details of your thinking. For example, other forest ecologists still haven't formed *your* hypothesis, designed and conducted *your* experiment, or collected, analyzed, and interpreted *your* data. Other law-enforcement officials haven't considered all the concerns brought to light by *your* committee about where to locate that new prison. So don't make too many assumptions about what your audience knows. What may be perfectly clear to you — because you've given it a protracted mental workout — may not be to someone coming upon it for the first time. However, where to draw the line between presenting too much information and not enough is tricky business that no book or instructor can generically prescribe: you'll have to think it through and use your discretion.

# Doing "Inner" Work

Much as a farmer must prepare the soil for planting, a writer must do the "inner" work of preparing the mental ground. Doing "inner" work begins with gathering your thoughts long before writing them down. Ideas stockpiled from previous work, backburner notions, information garnered from reading or synthesized from thinking — all need slack time to compost so they can fertilize and condition your mind to write.

To gather and compost effectively, focus on what you intend to write about, then "unfocus." That is, let your thoughts *gestate* just below the surface of consciousness. After all, technical writing is creative work, and creative work must gestate.

Gestation, once actuated, is unpredictable. Just when you're sure that nothing is happening, a great idea sprouts. So it's important to be prepared. To capture the fruits of ges-

tation, keep a little notebook handy so that as ideas occur —
and they will, as you're "unfocusing" while driving to work,
taking a shower, grocery shopping, or mowing the lawn —
you can jot them down while they're fresh, for later reference and elaboration. Realize, too, that gestation never com-
pletely stops till you're done writing. But its most critical role
comes early, before you've written the first word.

Alternately focus and unfocus for as long as you can: a few
hours if that's all you have (and it may well be in the real
world of short deadlines and impatient supervisors or cli-
ents), a few days (better yet), or (luxury!) a few weeks. The
longer the gestation period, the better jelled and ordered
your thoughts will be, and the better equipped you'll be to
begin writing.

# OVERCOMING WRITER'S BLOCK

*Did you turn to this chapter first?*

There's a fine point beyond which "gestation" becomes "procrastination." We're all familiar with this one. First, you brew a pot of coffee. You go to the bathroom. You make a few phone calls. You sort through old mail and papers on your desk; you take care of an invoice that should have been processed a week ago. Finally, you can't detour any longer. You must confront it: that report, that article, that book chapter, that letter, that grant proposal. And as you do, every coherent thought in your head flees. You're up against the wall that all writers face from time to time: writer's block. But how to get past it?

**First: Mark territory.** Designate a physical place you always go to to write, or adopt an attitude that prepares you to write in any physical place. The idea is to develop a conditioned response to this specific stimulus: the sight or idea of your writing "space" prepares you to write.

Seek privacy, which frees your ego from worrying about who's looking over your shoulder to see what a mess you've made. No one else should be, or should need to be, looking over your shoulder. To create, you can't afford to feel self-conscious. Indeed, to create, you need to literally get *out* of your self.

Remove yourself from peripheral chatter and traffic. Let your telephone answering machine take your calls. Turn off the stereo, TV, radio: tune in to the sounds within your own head.

Remove all other possible distractions — today's newspaper beckoning from the corner of your desk, a note prominently placed reminding you to send your mother flowers because it's her birthday. Remember: the mind that wanted to procrastinate in the first place is still, at some level, looking for an out.

**Next: Warm up by starting anywhere.** Don't worry about starting to write from the introduction and working sequentially through to the conclusion. Start with the *easier* parts, those that come more naturally to you, and leave the harder parts till later.

For example, in the typical scientific article, the introduction is a devil to write "cold" and is better left until the late stages of manuscript writing. Indeed, introductions composed without proper warm-up usually take a long running start and then ramble — and ultimately need to be rewritten. Likewise, the discussion section requires a thoroughly primed, smoothly humming brain to synthesize and interpret information. It shouldn't be tackled prematurely. Logic, if nothing else, dictates that the abstract, though literally the opening piece of text, should be written last; it is, after all, a capsule version of the larger document and should be distilled from it. In contrast, the methods section (in which procedures are outlined) and the results section (in which findings are presented without interpretation) are often good starting points because they're a little more straightforward. **Then: Grab any relevant thought and write it down.** Just put one word after another and, before you know it, you'll have sentences and (gracious!) paragraphs. Focus in the present — don't worry about the report summary while you're writing the objectives; don't think about how you'll illustrate the text before it's composed. But keep your mind limber, not *too* focused, so you continue to reap the fruits of gestation.

If pencil and paper make you freeze, try the typewriter or word processor. If writing machines make you freeze, try old-fashioned pencil and paper. If you're a facile speaker, try talking into a tape recorder or dictaphone, then transcribing your recording. Experiment: use what works best for you.

But whatever the process, try to relax so that your words flow naturally. Many people, uptight about how their prose "should sound," *write* things they would never *say*. For instance, why would you write

> The majority of the population relocated from rural to urban habitats.

when you would say

> Most people moved from the farm to the city.

You'll also find that a naturally flowing first draft will more readily yield to being revised later (see "Editing Your Own Writing," p. 103). A stilted first draft will tend to be less approachable, defying the editing assaults it badly needs.

**Above all: Keep the flow going.** Don't worry too much at first about sentence construction, word usage, eloquence, details of format, and so on. Plan to address those later while revising what you've initially composed. And don't worry too much yet about organization: the logic flow of your material will begin to take shape on its own. If you hit a snag in your thinking, insert a "Note to me" in the draft at the sticky point. Briefly describe what's needed, to cue yourself for later, and *move on.* Then, once the document's roughed out, you can return to your "notes" and address each in turn with full concentration and without the danger of losing your larger train of thought.

The idea here is to generate a mound of raw material to work with — something tangible, and therefore retrievable. Think of the rough text as a big, shapeless lump of clay, and of you the writer as a ceramic sculptor. Once the clay is before you, you can knead and mold it, flatten and rebuild it, until you're satisfied with the final product.

**Throughout: Tune into internal feedback.** Writing is a system: a set of parts that influence one another's behavior. And as in any system, all parts are not only connected but subject to feedback.

Think, for a moment, of another system you know well: the planet Earth. In tropical rainforests — one of Earth's subsystems — trees cycle immense amounts of water, and their roots network to hold fragile soils. If too many trees are cut down too fast, rainfall drops off as the water cycle is disrupted, and topsoil washes away as the mesh of tree roots deteriorates. That's "connection." But with the trees gone, the climate drier, and the soils impoverished, an environment that once supported forest can now support only grassland *because the act of cutting down the trees altered the ability of that environment to grow trees.* That's "feedback."

The concept of feedback — the process by which the factors that produce a result are themselves modified by that result — is crucial in writing. But how exactly does it operate?

Through "inner" work, you gather your thoughts; your thoughts gestate; then you write them down. That's "connection." But when you reread what you've written, you inevitably rethink and then rewrite some of it *because the think-*

*ing that produced the writing is itself modified by the writing.* That's "feedback."

This feedback loop — the repeated process of retreading traveled ground, of returning to knead and mold raw material — is a key aspect of any creative act. You should expect, and be grateful for, continuous feedback between thinking and writing throughout all stages of composing and revising a document.

# INVOKING A HELPFUL METAPHOR

Why not think about technical writing as something a little warmer and furrier than grammatical abstractions? Why not a vertebrate creature like yourself? And if writing is a vertebrate like you, by the same logic its body must resemble yours.

**Think of the skeleton of writing as *order*.** When text is orderly, one thought flows naturally and harmoniously to another, and elements similar in function are similar in form. When disarrayed, writing totters.

**Think of the body mass of writing as *conciseness*.** When text is concise, every word is wanted and needed, and ideas are rounded out rather than obscured. When bulky, writing lumbers.

**Think of the muscle tone of writing as *vigor*.** When text is vigorous, its lively nature compels attention. When flaccid, writing languishes.

The core of this book — the upcoming sections on order, conciseness, and vigor — carries through this metaphor, exploring the ailments that can befall the vertebrate creature of writing and illustrating how to keep the creature healthy. But conceptualizing writing as an organism whose body parts, like yours, must work together has an added benefit: it reinforces the key point that all system parts act in concert — never in isolation — to convey meaning and ensure the ultimate goal, clarity.

# Order —
# the Skeleton:
# Constructing a
# Stable
# Framework

# Logic Flow

You've defined your audience, allowed your thoughts to gestate, and begun priming your brain to write. Through these processes, you've identified several important points you want to cover and corralled ample supporting data to bolster each. But the best logic flow — that is, the "information hierarchy" of main and subsidiary points — is hard to settle on. Your dilemma, an impressive one, is how to transform a riotous mob of thoughts into a tidy processional.

## Basis for organization

Every document is built from a skeleton of main points (the more important ideas) and subsidiary points (the less important ideas). But how to identify which are which, and then arrange them so they cohere?

In this case, as in so many others, tackling the problem as a whole is far more difficult than tackling it piece by piece.

To begin, identify in your own mind the principle underlying your main points — that is, the basis for organization.

The basis forms the *backbone* of the skeleton of order. Some writers can identify the basis so naturally, so unconsciously, that they're unaware of the process involved. But most writers — who struggle with getting organized — need to consciously think through the process to mimic what doesn't come naturally.

There are many possible bases of organization. What, then, might some of them be?

- Perhaps the most obvious basis is *time*. For example, you might organize a report on economic trends during the 1950s and '60s on the basis of time (the chronology during a specified period).

- Reminiscent of *time* is *sequence*. For example, you'd probably organize a "how to" manual for car repair or an instruction booklet for operating a VCR on the basis of sequence (a series of steps in a prescribed order).

- Related to *sequence* is *progression*. For example, you might organize a book chapter on modeling long-term forest productivity on the basis of progression (general to specific) — first broadly defining the concepts of model and modeling, then presenting the various types of computer models appropriate to forestry, and finally naming a particular model to discuss in detail.

- Complementary to *time* is *space*. For example, you might organize a scientific article documenting the same set of experiments conducted in different locales on the basis of space (field sites).

- Often appropriate is *importance*. For example, you might organize an article on heart disease by discussing the various contributing factors (genetics, diet, smoking, and so on) in order of their perceived importance — probably beginning with the most important and proceeding to the least.

- Akin to *importance* is *concept*. For example, you'd likely organize a cookbook on the basis of concept — let's say, different parts of a meal (breads, desserts, soups, vegetables, appetizers, salads, main courses).

Organizing further, you could present the parts of the meal in the order in which they'd logically appear on the table (appetizers, soups, breads, salads, vegetables, main courses, desserts) — that is, you could *sequence* the *concepts* (use two bases of organization). Or you could simply present the parts of the meal in *alphabetical* order (a̲ppetizers, b̲reads, d̲esserts, m̲ain courses, s̲alads, s̲oups, v̲egetables), a handy neutral basis.

Some subjects dictate what the basis of organization will be. But many require that the writer decide. For instance, suppose you're planning an article on the activities of "underground" resistance organizations operating in Europe during World War II. Your basis could be *time* (chronology of events as they unfolded), *space* (country), *importance* (relative impact of each organization), or *concept* (political ideology of each).

# Categorizing information: the key to outlining

After identifying the basis for organization — the principle underlying your main points and the backbone of order — you need to array *the rest of the skeleton.*

Again, tackle the problem piece by piece. Begin by breaking the whole of your information into smaller, more manageable chunks. Then observe which ones group together naturally. By so doing, you'll begin to categorize information — which, as you'll soon see, is the key to outlining.

Categorizing is conceptually a two-part process:

- You *rank* information according to its relative importance. Typically, the main points form the major divisions preceded by first-order headings. The subsidiary points form the subdivisions preceded by lesser order headings.

- You *sequence* the ranked information — the divisions, and then within each division the subdivisions — by deciding which one is the natural opener, which should logically follow the opener, which should come next, and so on.

For example, I've ranked information in this book at three levels. Note, in the listings that directly follow, that the headings are not yet in any particular order because they are only being *ranked*, not *sequenced*.

My main points form *sections preceded by first-order headings* (in **boldface** type below):

**Technical Writing: the Agony & the Ecstasy**
**Order — the Skeleton: Constructing a Stable Framework**
**Prologue: Why This Book?**
**Starting: Before Writing the First Word**
And so on.

My subsidiary points form both *chapters preceded by second-order headings* (underlined below) —

Overcoming writer's block
Defining your audience
Logic flow
Continuity
Doing "inner" work

Invoking a helpful metaphor
Dynamics
Transitions

— and *subchapters preceded by third-order headings* (un-formatted below) —

Categorizing information: the key to outlining
Basis for organization

And so on.

Logically sequencing the sections, chapters, and sub-chapters (the ranked information) generates the current table of contents, an information hierarchy reflecting the main and subsidiary points:

**Prologue: Why This Book?**
**Technical Writing: the Agony & the Ecstasy**
**Starting: Before Writing the First Word**
Defining your audience
Doing "inner" work
Overcoming writer's block
Invoking a helpful metaphor
**Order — the Skeleton: Constructing a Stable Framework**
Logic flow
Basis for organization
Categorizing information: the key to outlining
Transitions
Dynamics
Continuity

And so on.

It's important to understand that the two-*part* process of ranking and sequencing is not a two-*step* process. That is, it appears linear but really isn't. True — you begin by *first* ranking and *then* sequencing the pieces of information. But the feedback that develops between thinking and writing as they both evolve means that you're soon also ranking and sequencing *simultaneously.*

It's equally important to understand that, in its early phase, categorizing relies heavily on trial and error. I didn't arrive at this book's organization on the first or even the second try. Trial and error shape your initial listing of ranked, sequenced categories. But — again — the feedback that develops prompts you to retread traveled ground and revise. So be prepared to alter the initial listing more than once, perhaps several times, maybe many.

Once you're satisfied with the ranked, sequenced listing of categories you've constructed, you'll have produced an *outline.* After all, an outline is just a stepwise structure for ordering information, the "bare bones" upon which to hang the flesh of exposition. And this outline will be as close as you can come to a "linear display" of the complexly related, nonlinear mental pattern of ideas you're trying to communicate.

But suppose you're blocked about how to start categorizing?

As with any kind of writer's block, start anywhere. Don't worry about starting at the beginning because you may not yet know where the true beginning is. Jot down as many key words and phrases as you can think of — perhaps on index cards or little slips of paper — so that all your information is retrievable. Don't worry at first about which word or phrase denotes a main point and which a subsidiary point. That may not yet be evident but will become so as you proceed. Move the cards or slips of paper around physically, like puzzle pieces, to see how they "flow" best. This way, you can try different outlines — that is, different ranked, sequenced listings of categories — and then decide which seems most natural to you.

Once you've established a satisfactory listing, notice how rough or refined it is. Is the outline a broad one from which you'll begin generating big, raw chunks of text? Or is it so detailed that you'll essentially be "filling in the blanks"? No one type of outline is best: experiment and see what works for you.

But remember that, whatever its configuration, your outline isn't "cast in stone." It's something flexible, pliable, a working framework still subject to the internal feedback that helped create it.

# TRANSITIONS

*O*nce you've arrayed the bones, you'll need transitions, *the connective tissue*, to gracefully link all the parts. A transition is most frequently a single word or phrase but may be an entire sentence or even a paragraph. It's the relationship between the linked parts, rather than the number of words that do the linking, that's at issue.

For instance, sense suggests that the following four sentences are related:

> The incumbent, from the well-to-do bedroom community of Walton, will probably be re-elected to office. Most of the large bloc of suburban voters favor him, according to the polls. Many inner-city voters support the challenger. He rose from the streets.

But the exact nature of the relationships isn't crystallized until the connective tissue of two transitions (**boldface** type) is added:

> The incumbent, from the well-to-do bedroom community of Walton, will probably be re-elected to office **because**, according to the polls, most of

the large bloc of suburban voters favor him. **Not surprisingly**, many inner-city voters support the challenger, who rose from the streets.

Notice that adding transitions also prompts the writer to reconstruct the sentences (see "Sentence Type," p. 83).

In the following example, the one-sentence paragraph (**boldface** type) functions as a vital transition:

> Once you're satisfied with the ranked, sequenced listing of categories you've constructed, you'll have produced an *outline*. After all, an outline is just a stepwise structure for ordering information, the "bare bones" upon which to hang the flesh of exposition. And this outline will be as close as you can come to a "linear display" of the complexly related, nonlinear mental pattern of ideas you're trying to communicate.
>
> **But suppose you're blocked about how to start categorizing?**
>
> As with any kind of writer's block, start anywhere. Don't worry about starting at the beginning because you may not yet know where the true beginning is...

Where transitions are absent, writing lacks both clarity and grace:

> The Douglas-fir and ponderosa pine seedlings were first planted in the study plots in May 1988 during warm weather. They were protected from deer browsing with paper budcaps. Deer were particularly abundant that spring. The seedlings were heavily browsed. Browsing obscures subsequent height differences, one of the key experimental variables. The study plots were replanted with seedlings of the same species in March 1989 during cold wet weather. The seedlings were encased in more durable plastic mesh tubes immediately after planting.

The combination of added transitions (**boldface** type) and varied sentence types makes the writing more focused and fluid:

> The Douglas-fir and ponderosa pine seedlings, first planted in the study plots in May 1988 during warm weather, were protected from deer browsing with paper budcaps. **However,** deer were particularly abundant that spring, and **despite protection**, seedlings were heavily browsed. **Because** browsing obscures subsequent height differences, one of the key experimental variables, the study plots were replanted with seedlings of the same species in March 1989 during cold wet weather. **This time**, the seedlings were encased in more durable plastic mesh tubes immediately after planting.

But writing that lacks transitions is more than fuzzy and awkward. It's dangerous. Not fully gestated, it requires the reader to "climb into the writer's head" to complete the job. That is, it leaves the reader to infer how one idea relates to the next. And if the reader's inference is not what the writer intended, meaning will — however inadvertently — be altered.

A writer should minimize the number of "decision points" for the reader. Ideally, there should be none. Danger lurks at every crossroads at which readers must guess which path they were meant to follow. The fewer decision points, the less likely a reader is to go astray.

# DYNAMICS

Writing, like music, depends on its dynamics — that is, its changes in energy levels — to cue the audience, through varying degrees of "loudness" and "softness," about when they'd better pay close attention and when they can afford to drift. Think about music: an entire piece played *fortissimo* or *pianissimo* will quickly lose its listeners because, with no counterplay of contrasts, with its passages all on the same dynamic level, it lacks focus.

You as a writer need to reinforce the logic flow you've established by making the main points "louder" and the subsidiary points "softer" to cue readers and keep writing focused.

Further, you need to plant your cues strategically. A key idea to understand here is that logic flow in writing differs from (the more familiar) logic flow in argument, in which you build "blind," point by point, from the bottom to the apex. Most readers will tire before ever reaching the apex unless there's foreshadowing of what's to come.

But how to keep readers interested and on track?

**You can selectively position information.** In writing, relative position dictates importance. Placing an idea at the beginning or end of a sentence, paragraph, chapter, or section *elevates* it in importance (makes it "louder"). Placing the same idea in the middle *diminishes* it (makes it "softer") and can effectively bury it.

For example, major findings from an experiment should boldly lead off the results section of a scientific article — after all, isn't this what the reader would want to know about first? Less conclusive or consequential findings can follow later. Similarly, key concepts in a marketing report should be highlighted in the executive summary up front, form the major sections of the report (further emphasized by first-order headings), and be recapped at the end — to linger in the reader's mind.

**You can delete information.** That's right, delete it! Graduate students writing theses, pay special attention here.

Don't try to tell readers *everything* you know about a given subject, just what they need to know (the "louder" points). Some of the more peripheral ("softer") points, though of interest to you, may not be relevant to the intended audience. And if they're not germane, they should be cut and stored for possible resurrection at another time. Some information may *never* find its way into print. The caveat here is, alas, a painful one to many writers: don't become blindly enamored of your material — it isn't precious.

# CONTINUITY

Sometimes, while you're in the heat of composing, continuity — that is, the coherence of the whole — goes haywire.

For instance, it's important to identify your objectives early, to cue readers about where you're headed, as in the following final paragraph of an introduction to a book chapter:

> In this chapter, we examine the stresses to which forest ecosystems currently are subjected; re-assess our traditional management approach, which emphasizes simplicity rather than complexity; and propose new or modified practices that better account for the uncertainty of these changing times.

However, this author laid out a set of chapter objectives and then failed to follow them in the ensuing text. Because she was "too close" to the material, she even failed to spot the discrepancy while revising.

Or you might unintentionally position text where it doesn't belong, perhaps in the wrong paragraph or under the wrong heading, as in the following excerpt from a methods section of a report:

**First-year measurements**

The plant population on each of the three sites is scheduled to be measured six times: at the beginning of the growing season (late spring), at the end of the growing season (late fall), and four times in between at equally spaced intervals. It is critical that the last measurements be made before snowfall renders these sites inaccessible. The other site visits will be in late fall of the second, third, fifth, and tenth years.

This writer lost sight of the heading ("First-year measurements") by the third sentence, in which he talked about measurements in years other than the first.

The above examples are strong evidence that writers, though "moving forward," must remember to "look back" to ensure continuity. The chapter author should have returned to the introduction — admittedly one of the harder parts to write — and revised it in light of what actually followed, *or* realigned chapter text with the originally stated objectives. The report writer should have moved the misplaced third sentence elsewhere, *or* changed the heading to accurately reflect the text that followed. Such self-monitoring is crucial when you're making order.

To foster continuity, it's an excellent practice to read over what you wrote during your last writing session just before beginning your next session. This helps to (1) reorient you within the document as a whole, (2) remind you of what you said last, and (3) prompt you about what you'll say next. Then, periodically during each writing session, stop actively writing and reread the previous several sentences or the previous paragraph or two. Better yet, return to the beginning of the current session's text and read forward to your stopping point.

Looking back will also help you detect other subtler problems of continuity. For example:

> Annual variation in peak stream temperatures is
> likely to vary by several degrees.

The bare bones (**boldface** type) of this sentence tell the story:

> Annual **variation** in peak stream temperatures **is**
> **likely to vary** by several degrees.

Surely this sentence was meant to read (bare bones again in **boldface**):

> Peak stream **temperatures are likely to vary** an-
> nually by several degrees.

Hastily moving forward, the writer forgot to look back.

Or how about the following internal memo to advertising-agency staff:

> Nice job on the presentation last Wednesday to
> Cosmo Inc. We got the account.
>
> First, our client wants us to survey the existing
> market to determine the variety of product needs.
> We'll have to discuss whether to query by mail or
> phone and who to hire to conduct the survey.
> We'll need a breakdown by geographical area,
> age, sex, income level of household, and profes-
> sion. Because time is of the essence, we'd better set
> up a schedule once we've determined the survey
> particulars.

Reread the first sentence in the second paragraph: "First, our client wants us to...." Using the word "First" in this position implies that the client has several other wants besides the survey. That is, "First" leads the reader to expect a sequence that never follows. Indeed, the sequence that does follow has nothing to do with the client's desires but instead reflects the actions that ad-agency staff will take regarding the survey. Perhaps, then, the memo should be rewritten as follows (proper sequencing highlighted in **boldface**):

Nice job on the presentation last Wednesday to Cosmo Inc. We got the account.

Our client wants us to survey the existing market to determine the variety of product needs. **First**, we'll have to discuss whether to query by mail or phone and who to hire to conduct the survey. **Next**, we'll need a breakdown by geographical area, age, sex, income level of household, and profession. **Then**, because time is of the essence, we'd better set up a schedule once we've determined the survey particulars.

Now "First" appropriately introduces the sequence completed by "Next" and "Then." (Notice also how "First," "Next," and "Then" operate as transitions, linking the intended actions.)

By repeatedly looking back as you move forward in writing, you'll be less apt to disturb continuity and create disorder (illustrated here), to inadvertently repeat text that doesn't warrant it (see "Repetition," p. 79), or to foster monotony (see "Sentence Type," p. 83).

# PARALLELISM

Visualize, for a moment, your own skeleton: a set of bones roughly symmetric around its center, the spine. If this arrangement were to go askew, you might limp, or one arm might dangle below your knee. More imaginatively, one leg might be connected to a shoulder instead of a hip.

Like your skeleton, the skeleton of the vertebrate creature of writing must have its parts symmetrically arrayed if it's not to totter. This means that parts which *function* alike should be *constructed* alike. That is, those parts should be parallel.

Parallelism is such a fundamental aspect of order that it ramifies through writing on three levels: (1) structure of an entire document, (2) structure of the internal parts of a document, and (3) structure within individual sentences. Let's look at examples at each level to see parallelism, or the lack of it, at work.

## Structure of an entire document

A good example of this grandest level of parallelism is a book's table of contents, which exposes the book's skeleton through a display of headings.

Most books have several levels of headings. First-order headings typically precede the main points, lesser order headings the subsidiary points. Because all headings of the same order function alike, they should be constructed alike. That is, they should be parallel.

To envision this, let's examine a broad excerpt of first- and second-order headings from this book's table of contents for parallelism (first order in **boldface**, second order <u>underlined</u>, and the core of the book further distinguished by *italics*):

> *Verb power*
> *Voice*
> *Grounded language*
> **Ending: After Writing the Last Word**
> Editing your own writing
> Improving your prowess as a writer

All first-order headings are parallel. Each leads off with a word or phrase representing a concept ("Prologue," "Technical Writing," and so on); that concept is followed by a colon (:), which introduces an elaborating phrase. Further, all first-order headings are formatted alike: they align left, and all words except articles ("the," "a") and conjunctions ("and") begin with a capital letter (see "Format," p. 65).

Now look closer, and you'll see parallel subsets of first-order headings nested within the larger grouping. Notice how the first-order headings from the book's core are parallel:

> ***Order — the Skeleton: Constructing a Stable Framework***
> ***Conciseness — the Body Mass: Making Every Word Count***
> ***Vigor — the Muscle Tone: Empowering Your Words***

Each heading leads off with a single noun representing a concept ("Order," "Conciseness," "Vigor"); that noun is followed by a dash (—) and then the appropriate body element of the vertebrate creature of writing ("the Skeleton," "the Body Mass," "the Muscle Tone"). Each body element is followed by a colon, which introduces a gerund[1] phrase ("Constructing a Stable Framework," "Making Every Word Count," "Empowering Your Words"). Further, all headings are formatted alike: they align left, and all words except articles begin with capital letters.

Notice also how the book's core (here abbreviated) is bracketed by parallel section headings (<u>underlined</u>):

> **Starting: Before Writing the First Word**
> *Order:...*
> *Conciseness:...*
> *Vigor:...*
> **Ending: After Writing the Last Word**

This adds yet another element of symmetry to the book's skeleton.

The second-order headings, though all formatted alike, also illustrate a more complex structure: those within the book's core are parallel, and those in the sections bracketing the book's core are parallel. That is, each second-order heading from the core ("Logic flow," "Transitions," and so on) is one or two words representing a concept, whereas each heading from the sections bracketing the core ("Defining your audience," "Doing 'inner' work," and so on) is a gerund phrase. Here *the need for emphasis outweighs the need for strict parallelism.*

# Structure of the internal parts of a document

Good examples of this level of parallelism — intermediate between that for an entire document and that within individual sentences — are lists and descriptions.

Lists are simple arrays of items either embedded in the text or displayed (that is, broken out of the text). Because items in each array function alike, they should be constructed alike. That is, they should be parallel.

In the following example, a set of instructions for using an electric typewriter, notice how misaligned parts can hide more easily in an embedded than a displayed list (parallel constructions highlighted in *italics*):

**Embedded list**

Not parallel (sentences giving instructions are constructed differently):

The instructions are as follows. First, turn the power on. The next thing is that a sheet of paper is inserted. The margins and tabs should then be set, and the line spacing checked. Finally, you can begin typing.

Parallel (sentences giving instructions are constructed alike):

The instructions are as follows. *First, turn* the power on. *Next, insert* a sheet of paper. *Then, set* the margins and tabs *and check* the line spacing. *Finally, begin* typing.

**Displayed list**

Not parallel (sentences giving instructions are constructed differently):

The instructions are as follows:
(1) Turn the power on.
(2) A sheet of paper is inserted.
(3) The margins and tabs should be set, and the line spacing checked.
(4) You can begin typing.

Parallel (sentences giving instructions are constructed alike):

The instructions are as follows:
(1) *Turn* the power on.
(2) *Insert* a sheet of paper.
(3) *Set* the tabs and margins *and check* the line spacing.
(4) *Begin* typing.

Descriptions are more elaborate arrays of functionally similar items that should be constructed parallel. For example, if you're a biologist writing up a field study in which several sites are compared, you should establish a particular order ("order of mention") in which to describe site characteristics — elevation, annual rainfall, vegetation, soils, and so on — and then repeat that order for each site. Using the same order creates a mental template that allows easy comparison of similarities and accentuates differences.

Or suppose you're a wood-decay specialist writing up the results of a laboratory study in which lumber impregnated with each of four preservatives is examined for evidence of rot. You might name the four preservatives in your introduction, thereby establishing an order of mention (let's call it A, B, C, and D) that should be paralleled in the methods and results sections *unless the need for emphasis outweighs the need for parallelism.*

Recall, from "Dynamics" (p. 43), the value of making the main points "louder" and the subsidiary ones "softer" to keep writing focused and readers on track. So if preservative C is found to be the most effective at preventing rot and preservative A the least, with B and D falling in between, you should probably mention the preservatives "out of order" in the results — that is, C, B, D, and A (most to least effective) or C, A, B, and D (most and least effective, then intermediate). Even determining the best "out of order" order requires that you think it through.

# Structure within individual sentences

At this most localized structural level, parallelism operates through individual words and phrases to ensure that functionally similar elements are aligned.

For example, in the sentence

Seedlings will be planted to stabilize streambanks,

to provide forage for wildlife, and to improve aesthetics.

all three "purposes for planting seedlings" are constructed the same way — as infinitive[2] phrases (complete phrase in **boldface**, infinitive <u>underlined</u>):

Seedlings will be planted **to <u>stabilize</u> streambanks, <u>to provide</u> forage for wildlife, and <u>to improve</u> aesthetics.**

However, misalignment within a sentence can be subtle, often because the bare bones are obscured. Here *italics* highlights the problem:

**Not parallel**

The purpose of this little book is as much *to show* you the artistry of technical writing as *helping* you with the underlying mechanics.

**Parallel**

The purpose of this little book is as much *to show* you the artistry of technical writing as *to help* you with the underlying mechanics.

Or

The purpose of this little book is as much *showing* you the artistry of technical writing as *helping* you with the underlying mechanics.

The two aspects of "the book's purpose" should be parallel: alike in function, they should be alike in construction, either both infinitives ("to show," "to help") or both gerunds ("showing," "helping").

Any time you suspect that a sentence is internally misaligned — sometimes it just doesn't sound right — investigate to see what's gone awry. First, get down to the bare bones. Recall, from "Continuity" (p. 47), how stripping the flesh (<u>underlined</u>) from the bones (**boldface**) helped reveal the problem: the subject ("variation") and predicate ("is likely to vary")[3] were inadvertently repeated because the writer forgot to look back:

> <u>Annual</u> **variation** <u>in  peak stream temperatures</u>
> **is likely to vary** <u>by several degrees.</u>

The idea is to work piece by piece to isolate the offending text from its surroundings, which can neatly mask the trouble.

Key phrases that echo (repeat periodically throughout a document) also should be parallel. For instance, a doctor discussing the benefits to aerobic fitness of "walking, running, biking, and swimming" should continue to mention those activities in that order unless, of course, the need for emphasis outweighs the need for parallelism.

A word phrase operates much like a musical phrase: it has a certain "sound" to which the reader's ear becomes attuned. If you change the order of the words, you effectively change the "sound" — which may falsely signal to your reader a change in emphasis or meaning.

# HIDDEN DANGERS

Within sentences, dangers lurk beyond those that directly violate the precepts of parallelism (see the previous chapter). Even though hidden dangers are localized, they nevertheless deform the skeleton. The following are the three worst offenders.

## Skewed comparisons

One of the most insidious hidden dangers is the skewed (or false) comparison, which twists the meaning of a sentence by twisting its form. For example, inspect the following:

> My idea of what constitutes good writing may differ from my colleague.

Isolating the true comparison reveals that, structurally, it's between "idea" and "colleague" (**boldface** for emphasis):

> **My idea** of what constitutes good writing **may differ from my colleague.**

Surely, the writer intended otherwise:

> **My idea** of what constitutes good writing **may differ from** <u>that of</u> my colleague.

(My idea...may differ from that [ = the idea] of my colleague.)

Or

**My idea** of what constitutes good writing **may differ from my colleague's.**

(My idea...may differ from my colleague's [idea].)

Wherever text compares elements, look closely to be sure the comparison's a true one.

# Dangling phrases

The dangling phrase: much maligned, and rightly so! This little number can, and does, worm its way into the best of writing. Despite examples and explanation, danglers are hard to spot, so I give them disproportionate emphasis here.

A dangler is a structural misfit — a group of words (typically a participial[4] or an infinitive phrase) that connects illogically to the rest of the sentence. To really understand what a dangler is, you'll need to examine how it operates. So look closely at the following example (dangler in **boldface** type):

> **Once diagnosed as the problem**, the farmer must rely on existing information from the local extension agent and years of experience working the land to eradicate the microscopic soil fungus.

Sense tells you that the "problem" is the fungus, invisible to the unaided eye, which is rotting the farmer's crops. But sentence construction conveys something quite different: it tells you that the farmer is the "problem" because the phrase "Once diagnosed as the problem" structurally *hangs* — that is, dangles — from "farmer" (**boldface** for emphasis):

> **Once diagnosed as the problem, the farmer** must rely on...

To create the right logical connection, you must detach the past participle "diagnosed" from the grammatical subject of the sentence, "farmer," and associate it with its *true* subject, "fungus" (**boldface** for emphasis):

> Once the microscopic soil **fungus has been diagnosed** as the problem, the farmer must rely on existing information from the local extension agent and years of experience working the land to eradicate it.

The words "fungus" and "diagnosed" are now properly related as subject and predicate.

A second example reveals the present participle "Reviewing" doing much the same kind of structural dirtywork as the past participle "diagnosed" (dangler in **boldface**):

> **Reviewing the enclosed police report**, little evidence was found to support the allegation that my client, the accused, assaulted the plaintiff.

Inference tells you that the accused person's lawyer was the one reviewing the police report. But "the lawyer," the true subject of the participle "Reviewing," is only *implied* by the sentence, not included in it. To create the right logical connection, you might recast the previous sentence as follows:

> Reviewing the enclosed police report, I found little evidence to support the allegation that my client, the accused, assaulted the plaintiff.

Inserting the pronoun "I," a referent to "lawyer" — that is, the *doer* (agent) of the action in the sentence — instantly clarifies the muddy waters. First, it removes inference. (Recall, from "Transitions," p. 41, the danger of readers inferring what the writer meant to say but didn't.) Second, it properly aligns sense and structure. Finally, it gives readers more information (we are now told who did the reviewing) and invigorates the sentence by transforming it from passive to active voice. (See "Voice," p. 91, for additional examples of the frequent benefits of using active voice.)

You might have recast as follows:

> Little evidence was found in the review of the enclosed police report to support the allegation that my client, the accused, assaulted the plaintiff.

This version corrects the dangler through a different strategy. It converts the verbal "reviewing" to its noun form, "review." And eliminating the verb eliminates any need for a subject. But this passive recasting isn't as crisp as the active one with the inserted "I." And it's less informative: we still aren't told who did the reviewing, though, structurally, we don't need to be.

Note also that the converted phrase "in the review of the enclosed police report" has been moved midsentence simply because it sounds better there. It's the shift from the verbal to the noun, not the shift in sentence position, that eliminates the dangler.

Although a dangler is more prominent at the beginning of a sentence simply by virtue of position, it can crop up anywhere. For instance (dangler in **boldface**):

> Little evidence was found **reviewing the enclosed police report** to support the allegation that my client, the accused, assaulted the plaintiff.

In this variation, because the dangler is buried midsentence, it's less likely to jar the ear—and therefore more likely to be able to hide there unexposed. But the doer is only implied, not present, in the sentence, and the phrase continues to dangle.

A really hasty writer might even generate this variation:

> Little evidence was found to support the allegation that my client, the accused, assaulted the plaintiff, **reviewing the enclosed police report**.

The location of the dangler here does more than muddy the waters. It implies wrong meaning: hanging at the end of the sentence, it suggests by its position that the plaintiff, rather than the (still implied) lawyer, was doing the reviewing.

See if you can expose the dangler (**boldface** type) before I do in each of the following four numbered sentences and then make the right logical connection:

> 1 When comparing operational aspects of the shelterwood and clearcut methods for regenerating forest stands, five important factors must be considered.

**Dangler** (present participial phrase) exposed:

> **When comparing operational aspects of the shelterwood and clearcut methods for regenerating forest stands,** five important factors must be considered.
>
> (**When comparing...** dangles from "factors." The "factors" don't do the "comparing"; the land owners or managers do.)

Recast versions eliminating the dangler: [5]

> When comparing operational aspects of the shelterwood and clearcut methods for regenerating forest stands, land managers must consider five important factors.

Or

> When land managers compare operational aspects of the shelterwood and clearcut methods for regenerating forest stands, they must consider five important factors.

Or

> When operational aspects of the shelterwood and clearcut methods for regenerating forest stands are compared, five important factors must be considered.

◊

> 2 The music had to be turned up so loud to really feel like dancing that I could hardly hear my partner speak.

**Dangler** (infinitive phrase) exposed:

> The music had to be turned up so loud **to really feel like dancing** that I could hardly hear my partner speak.
>
> (**to really feel...** dangles from "music." The "music" doesn't do the "feeling"; the partners do.)

Recast version eliminating the dangler:

> To really feel like dancing, we had to turn up the music so loud that I could hardly hear my partner speak.

◊

3　Broken sometime during the night, we guessed from the blood on the floor that the burglar had been cut on the jagged glass protruding from the window.

**Dangler** (past participial phrase) exposed:

> **Broken sometime during the night,** we guessed from the blood on the floor that the burglar had been cut on the jagged glass protruding from the window.
>
> (**Broken...** dangles from "we." "We" weren't "broken"; the window was.)

Recast version eliminating the dangler:

> We guessed, from the blood on the floor, that the burglar had been cut on the jagged glass protruding from the window that he broke sometime during the night.

◊

4　When I was a small child, the concrete canyons between highrise apartments were my playground. At eight, my parents bid farewell to the city and bought a house in the suburbs.

**Dangler** (prepositional phrase) exposed:

> When I was a small child, the concrete canyons between highrise apartments were my playground. **At eight,** my parents bid farewell to the city and bought a house in the suburbs.
>
> (**At eight** dangles from "parents." The "parents" weren't "eight"; the narrator was.)

Recast version eliminating the dangler:

> When I was a small child, the concrete canyons between highrise apartments were my playground. But when I turned eight, my parents bid farewell to the city and bought a house in the suburbs.

# Mismatched subjects and predicates

Structurally less devious than either the skewed comparison or dangling phrase, but a hidden danger nevertheless, are the mismatched subject and predicate. A subject and a predicate are "mismatched" when one is singular and the other plural — the mismatch creating a sentence much like a mathematical equation that's unbalanced.

At first glance, mismatching subjects and predicates might seem less a hidden danger than a beginner's mistake. And it is, where the sentence is uncomplicated and its bare bones are readily apparent. For instance, it's hard to imagine anyone writing the following sentence containing the jarring **boldface** clause in which subject and predicate are mismatched:

> **Fine roots is easily stripped** when conifer seedlings are carelessly lifted from nursery beds.

But where modifying words and phrases obscure the bare bones, beware:

> The stripping of roots during lifting and the pruning of roots before packing physically reduces root volume of nursery seedlings.

Here, the writer mistakenly supplied the singular verb "reduces," probably because the singular "packing" is closest to the verb and looks, at a glance, like its subject. But exposing the bones, you'll discover that the true subject of this sentence, "stripping...and pruning" (two gerunds connected by a conjunction), is plural and properly takes the plural verb "reduce."

Writers sometimes become so intimate with certain technical phrases in their fields — especially coordinate pairs like "research and development" — that, unconsciously, they begin to think of these pairs as units. And as a result, the units (**boldface** type) are often "logically" but incorrectly assigned singular verbs:

> Once seedlings have been planted in the field, their **growth and survival** largely depends on elements out of the control of even the most diligent land manager.

In fact, the nouns "growth" and "survival" — though going hand in glove in the writer's mind — form a plural subject (like "stripping...and pruning" in the previous example), and this plural subject requires the plural verb "depend."

There are, of course, instances in which a pair of nouns *connoting a single concept* takes a singular verb:

> Cops and robbers was one of my favorite childhood games.

The point is that you must remain alert to the subtleties of sentence construction and word usage to ensure the right match.

# CONSISTENCY

$O$nce your writing is well constructed down deep, you can put the finishing touches on order by being uniform — that is, consistent — in your format and terminology.

## Format

Formatting is more than the mechanical detail work that makes your final product look its most presentable. It's the "window dressing" that reinforces deep structure.

For example, elements that are structurally parallel should be formatted consistently:

- **Headings** of the same order should look the same — All capital letters? Mixed capitals and lowercase? Centered between the left and right margins? Flush with the left margin? **Boldface?** *Italic?* <u>Underlined?</u> And so on.

- **Entries in a displayed listing** (like this one) should look the same — Each entry preceded by a number, letter, asterisk, or (as here) "bullet"? First word of each entry capitalized or lowercase? The end of each entry punctuated? If so, with comma, semicolon, period, or question mark? Spacing between entries? Spacing between the main body of the text and the listing? And so on.

- **Tables** should look the same — Table heading centered over the table or flush with the left margin? Heading in all capital letters or mixed capitals and lowercase? Heading pithy, with details in footnotes? Or comprehensive, to avoid footnotes? Horizontal lines ("rules") above and below table body? If so, how many? Vertical rules allowed within table? Rules or white space used to separate groupings within the body of the table? Units of measurement, if any, spelled out or abbreviated? And so on.

Format is usually specified by the person or organization you're writing for: a professional society (for a technical journal), a publisher (for a magazine article or book), an agency (for a commissioned report or grant proposal), a graduate school (for a thesis). However, format specifications may range from minimal (double-spaced copy) to detailed (a booklet containing a comprehensive set of instructions).

If specs are minimal or literally nonexistent, settle on some format conventions of your own. Jot them down (so you don't have to try to remember them all) as a project-specific style guide to keep handy while you're writing. It's a good idea to scan the guide to refresh yourself on details of these conventions before each writing session.

But because there's no generic "ideal format," the one you choose should be tailored to your material. For instance, if you're documenting a computer model that includes many lengthy displayed equations, don't select a snappy three-column magazine-type layout that will force you to break equations awkwardly. Save the more visually dynamic format for your article and accompanying color photos, targeted for a "pop" nature magazine, on spending the summer hiking the Pacific Crest trail.

# Terminology

Using inconsistent terminology is one of the worst traps in technical writing — and one of the easiest to avoid once you're sensitized to it.

Simply put: you should minimize the number of *different* technical terms you use to refer to the *same* thing. When terms "shift" like this, your readers, even specialists in your field, will have to infer what you meant. And, once again, wrong inference will lead to wrong meaning.

For example, if you're a field biologist, don't also refer to your "study plots" as "treatment areas," "sites," or "quadrats." Even the most astute reader will be forced to guess whether these four terms all refer to the same thing, whether some are the same and others are different (perhaps subunits), or whether all are different. You as the writer know because it's your study. But the reader can only divine.

Or if you've experimentally thinned a forest stand to three stocking levels, don't then proceed to discuss "the heavily, moderately, and lightly thinned plots" *and* "the low-, medium-, and high-density plots." Even though the two sets of terms are equivalents, using both causes the reader to have to mentally convert one set to the other every time you shift. Instead, state the equivalence between the sets at the first possible opportunity and then use just one set, whichever you prefer, thereafter:

> After fertilization, eight plots were heavily thinned (low density, 120 trees/acre), eight mod-

> erately thinned (medium density, 445 trees/acre),
> and eight lightly thinned (high density, 1475
> trees/acre).

Occasionally, you may have to use more than one term or set of terms. But try limiting yourself to two, and state right away the equivalence between the term you'll prefer thereafter and any alternates.

Shifters will insidiously insert themselves into your initial draft. So hunt them down as you revise. And remember to cross-check text, including headings and captions, with tables, charts, graphs, and illustrations, if you have any, to ensure consistency of terms among all aspects of your document.

However, the situation is the converse for *common* terms. Whereas it's best to use just one technical term to avoid reader confusion and wasted motion, it's generally desirable to vary common words or phrases to add spice to writing. For example, in a literature review, at its best never scintillating reading, you should prefer (common terms underlined)

> Although Smith and Jones (1977) reported that
> most of the evidence comes from short-term stud-
> ies, Brown (1985) notes an exception, the ongo-
> ing work, begun some 25 years ago, of Thomas et
> al. (1975, 1978, 1983). They found that...

to the soporific

> Although Smith and Jones (1977) found that
> most of the evidence comes from short-term stud-
> ies, Brown (1985) finds an exception, the ongoing
> work, begun some 25 years ago, of Thomas et al.
> (1975, 1978, 1983). They found that...

Indeed, varying common terms is an instance in which *diversity*, rather than uniformity, is the goal (see "Sentence Type," p. 83).

# Conciseness — the Body Mass: Making Every Word Count

# INFORMATION LOAD

How many words constitute a document's "ideal mass"? And how many more make it "wordy"? Is it only the number of words you must be concerned with — or is there something more?

At first glance, wordiness appears to simply be using *many* words to say what you mean. But it's actually using *too many* — more words than are needed to advance your thought. On this basis, a short sentence can be wordy, or a long one concise.

Wordiness, then, is not so much a function of quantity of words as of density,[6] or "information load" — that is, how much meaning words carry. Wherever words, whether many or few, carry too slight a load or virtually no load at all, they are "empty." And wherever too many words are empty, writing is wordy.

You can expect your early drafts, especially the first, to be wordy. That's when you're composing — when your main concern is to loose a flow of thoughts and then capture that flow while it lasts. But when you're revising, you should program yourself to examine words, individually and in groups, for information load, then dispose of the "empties" — which, like empty calories, add only bulk — and rewrite.

Eliminating the empties usually involves reconstruction, not just simple deletion. For example, you might generate a flood in the heat of writing your congressional representative about a sensitive issue:

> Dear Congressman Avariss,
>
> No one would argue that we need to put serious effort behind developing sources of energy other than oil and coal, which we've relied on for so long now. But I have long feared the development of nuclear power as an alternative energy source because no progress in safe disposal of the radioactive wastes produced by nuclear power plants has yet been made. Many people seem to be of the opinion that once the wastes are buried or dumped in the ocean, we can simply forget about them.
>
> But for me the matter has suddenly taken on new dimensions. I read in yesterday's *Times-Herald*, our local newspaper, that you're supporting possible location of a nuclear waste disposal site only 40 miles southeast of here. I also read recently, in the same paper, that you own a large tract of land in what appears to be that same vicinity. What a coincidence that the government might want to buy land — and for a handsome price, no doubt — so that it can locate its waste facility in an area where you profit from selling your own substantial acreage....

But you'll need to mop up the excess once you've cooled down:

> Dear Congressman Avariss,
>
> Granted: we need to develop sources of energy other than oil and coal. But I have long feared development of nuclear power as an alternative energy source because the problem of safely disposing of the radioactive wastes hasn't been

satisfactorily resolved. The reigning philosophy
has seemed to be "out of sight, out of mind."

But for me the matter now has new dimensions. I
read in yesterday's *Times-Herald*, our local newspaper, that you're supporting possible location of
a nuclear waste disposal site only 40 miles south-
east of here. I've also recently learned that you
own a large tract of land in that vicinity. What a
coincidence that the government might want to
locate its waste facility where you own substantial
acreage and stand to profit....

Or you might effervesce in an early draft about the computer model you've developed to project national supply and
demand of lumber commodities:

The Timber Commodity Evaluation Model
(TCEM) was developed for use in projecting future
supply and demand of stumpage, lumber, and
secondary wood products from U. S. national for-
ests. Early in the development of this model, it be-
came apparent that there would be a need to
simulate the effects of management intensification on private lands. Timber harvest projections
from the nation's public forest lands are not calculated by TCEM but instead include assumptions
about management intensification; these assumptions about public harvest are fixed by annual allowable sale quantity calculated in the
management plans for individual units of pub-
licly owned forest land. To effectively project an
increase in management intensification on private lands, it was necessary to adjust the growth
and mortality constraints within the model in a
fashion that would reflect the effects of increased
management intensity.

The major reason why this adjustment was
needed was that the investments in timber management can alter future timber supplies. For example, an increase in management intensification on forested acres now such as precommercial

> thinning could have a substantial impact upon the amount of inventory volume in the future. A large amount of inventory volume available would cause a greater harvest volume and a lower price in future time periods. It was the capability to address this type of policy analysis that was needed. In order to account for these effects, we constructed the Timber Analysis Program (TAP), a subroutine that embodies a concerted effort to reflect growth dynamics of the stand and maintain the biological integrity of the inventory unit....

But you'll need to degas when revising:

> The Timber Commodity Evaluation Model (TCEM) was developed to project supply and demand of stumpage, lumber, and secondary wood products from U.S. national forests. However, early in model development, the need to simulate the effects of management intensification on private lands became apparent. (Management intensification on public lands, not calculated by TCEM, is instead accounted for through assumptions fixed by the annual allowable sale quantities calculated in public management plans.) To project for private lands, we had to adjust TCEM's growth and mortality constraints to accommodate investments in timber management, which can alter future timber supplies; for example, managing more intensively now by precommercial thinning could substantially affect future inventory, increasing harvest volume and lowering prices. To this end, we constructed the Timber Analysis Program (TAP), a subroutine intended to reflect stand growth dynamics and maintain the biological integrity of each inventory unit....

It's possible to eliminate too many empties, to make writing *too* dense and without enough "breathing space." But this problem is relatively rare and will not be dealt with here.

# OVERLAP

Wordy writing is usually rife with overlap — that is, sets of words which echo one another to no purpose. Too much overlap in writing gives readers the feeling they are "running in place," or, at best, making painstakingly slow forward progress. For example,

> A contract has been awarded to two forest-tree nurseries. The nurseries are Wonder Plant Nursery in Oregon and Bigtree Nursery in Washington. These nurseries will grow seedlings from seed collected from local zones. The resulting seedlings will be ready in November. They will be planted to stabilize streambanks, to provide forage for wildlife, and to improve aesthetics.

The first clue here about overlap is that the word "nurseries" appears in each of the first three sentences and is the subject of two of them. The second clue is that all five sentences are the same sentence type (simple sentence: a single independent clause).

But you can eliminate overlap by

- Examining individual words and word structures (sentences, clauses, and phrases) for their proper information load,

- Compacting elements that carry a slighter load to slighter structures — a sentence to a clause or phrase, a clause to a phrase, a phrase to a word — and then

- Nesting the slighter structures within the grander ones.

Grammarians call this compacting and nesting process "subordination." And using subordination, you can improve the previous example to make it more concise:

> A contract has been awarded to two forest-tree nurseries, Wonder Plant Nursery in Oregon and Bigtree Nursery in Washington, which will grow seedlings from seed collected from local zones. The resulting seedlings, to be ready in November, will be planted to stabilize streambanks, to provide forage for wildlife, and to improve aesthetics.

Let's look at some of the improvements. The second sentence of the original version —

> The nurseries are Wonder Plant Nursery in Oregon and Bigtree Nursery in Washington

— has been compacted to a phrase:

> ~~The nurseries are~~ Wonder Plant Nursery in Oregon and Bigtree Nursery in Washington

The third sentence of the original —

> These nurseries will grow seedlings from seed collected from local zones.

— has been compacted to a clause:

> ~~These nurseries~~ [which] will grow seedlings from seed collected from local zones

- And both the phrase (<u>underlined</u>) and the clause (**boldface**) — now slighter structures than they originally were as sentences — have been nested within the first sentence of the original:

> A contract has been awarded to two forest-tree nurseries, <u>Wonder Plant Nursery</u> <u>in Oregon and</u> <u>Bigtree Nursery</u> <u>in Washington</u>, **which will grow seedlings from seed collected from local zones.**

The fourth and fifth sentences of the original version also have been similarly compacted and nested.

Because word structures now carry their proper load, the revised version is more concise. And because sentence types were diversified as a by-product of subordination (see "Sentence Type," p. 83), the revised version is more vigorous and fluid than the original, which was choppy from the monotony of only one sentence type and no transitions between sentences. To fully feel the difference, read the two examples aloud back to back.

# REPETITION

You've probably noticed that removing overlap in writing involves removing text that has been repeated unnecessarily. To be sure, there are times when you *want* repetition to emphasize an important point. But most repetition is inadvertent and only nourishes wordiness.

Examine the following wordy sentence, excerpted from a research report, for needless repetition and the poor information loading often associated with it:

> Interpretation of the data collected to this point is somewhat limited because information has been collected only for this first year of the study.

The clues here about repetition are that the word "collected" appears twice and that the words "data" and "information" seem to refer to the same thing. However, by eliminating needless repetition, you can condense and clarify this bleary sentence:

> The data collected so far is only for one year; therefore, its interpretation is limited.

Or how about the following:

> Stream temperature measurements will be made
> during late spring and summer, the season of
> maximum leaf area and maximum solar radia-
> tion. Measurements will be taken for each of
> seven years, with measurements continuing be-
> yond that as funding permits.

The clues here are that "measurements" is repeated three times in two sentences and that "maximum" is repeated twice as a modifier of related nouns. This bulky text might be rewritten:

> Stream temperatures will be measured once
> yearly for seven years (and thereafter as funding
> permits) during late spring and summer, the sea-
> son of maximum leaf area and solar radiation.

Notice also in this example how the writer reduced overlap between the two original sentences, compressing them into one, to eliminate repetition.

Remember that repetitious words or phrases physically close to one another — within the same sentence or paragraph, like the two previous examples — are easier to spot and weed out than repetitious passages separated by multiple pages. So keep a sharp lookout for *hidden* repetition as you periodically look back while moving forward (recall "Continuity," p. 45) and when you're revising a draft (see "Editing Your Own Writing," p. 103). For example, be sure that the statistical details of the Harvey Wallbanger Multirange Test, described in the methods section of your article, aren't recapped in the results section. Be sure that the findings about the effectiveness of preservatives A, B, C, and D, reported factually in the results section, don't form a running start for the discussion. And be sure that your concerns about nuclear waste disposal, the springboard to the real issue — graft — prompting the irate letter to your congressman, open the letter but aren't resurrected in later paragraphs.

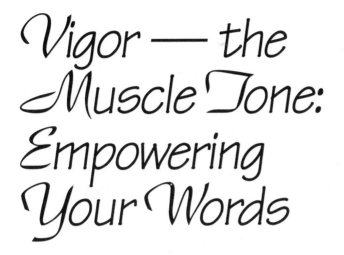

# Vigor — the Muscle Tone: Empowering Your Words

# SENTENCE TYPE

You as a writer can further manipulate dynamics, which cues readers about the importance of a passage and forestalls monotony, by varying sentence type.

Grammarians name sentence types to reflect their construction. The names themselves are not important — they're just handy labels. What is important is that you recognize the differences in construction to ensure a rich mix of sentence types to keep writing lively. Here again, *diversity*, rather than uniformity, is the goal. The explanation accompanying each of the following sample sentence types should help elucidate the differences:

*It is raining today.*

> **Simple sentence**: one independent clause. Recall from English class long ago that a clause must have a subject (in this case, "It") and a predicate (in this case, "is raining"), and that an independent clause is so named because it can stand alone as a sentence.

*It is raining today, but the forecast is for sunshine tomorrow.*

> **Compound sentence:** two independent clauses ("It is raining today" and "the forecast is for sunshine tomorrow") linked with the conjunction "but." Note that if "but" were removed, each of the clauses could stand alone as a sentence ("It is raining today" and "The forecast is for sunshine tomorrow").

*Although the forecast is for sunshine tomorrow, I'll believe it when I see it.*

> **Complex sentence:** one independent clause ("I'll believe it") and one or more (in this case, two) dependent clauses ("Although the forecast is for sunshine tomorrow" and "when I see it"). Recall from English class that a dependent clause is so named because it cannot stand alone as a sentence — that is, it depends on other structural elements. Dependent clauses are introduced by words such as "although," "when," "because," "since," "that," "which," or "who."

*It is raining today, but although the forecast is for sunshine tomorrow, I'll believe it when I see it.*

> **Compound-complex sentence:** the sample compound sentence above combined with the sample complex sentence — that is, more than one independent clause with at least one dependent clause.

Of the four sentence types, the simple sentence is the type most likely to be overused and, as a result, to create monotony, as is evident in the previous example from "Overlap":

> A contract has been awarded to two forest-tree nurseries. The nurseries are Wonder Plant Nursery in Oregon and Bigtree Nursery in Washington. These nurseries will grow seedlings from seed collected from local zones. The resulting seedlings

will be ready in November. They will be planted to stabilize streambanks, to provide forage for wildlife, and to improve aesthetics.

Recall how this text was invigorated by reconfiguring the first three simple sentences into one complex sentence (**boldface** type):

> **A contract has been awarded to two forest-tree nurseries, Wonder Plant Nursery in Oregon and Bigtree Nursery in Washington, which will grow seedlings from seed collected from local zones.** The resulting seedlings, to be ready in November, will be planted to stabilize streambanks, to provide forage for wildlife, and to improve aesthetics.

Notice also in the sample sentence types that as sentences become more intricate in structure, they tend to become longer. But this doesn't follow automatically. For instance, a highly descriptive simple sentence may be far longer than a spare complex sentence:

> **Simple sentence**
>
> Nursery management in the Pacific Northwest today represents the greatest concentration of technology and investment in the forest growth cycle and rivals wood processing in capital and labor intensity.

> **Complex sentence**
>
> As harvest levels increase, the demand for nursery stock also will increase.

In the simple sentence, the subject ("management") and predicate ("represents...and rivals") are elaborated with many prepositional phrases. In the complex sentence, the independent clause ("the demand for nursery stock also will increase") and dependent clause ("As harvest levels increase") make a terse statement.

No matter how stimulating your content may be, too many sentences of the same type, the same length, or some combination of the two make for deadly dull writing. (See "Voice," p. 91, for additional ways of varying sentence type.)

# VERB POWER

Verbs are the words that carry the action of the sentence. Okay: you've heard that one before.

But they're more than that: they're the power words. And their power energizes writing best when verbs are allowed to shoulder most of the information load.

Recall that text carrying too slight a load or virtually no load at all makes writing *wordy*. Here notice how too slight a load adds insult to injury by making writing *flabby* as well. Consider the following excerpt from a stream biologist's report:

> An examination of the trout in this stream for parasites will be done during the summer of 1992. Initial data analysis and interpretation will occur at that time.

Much of the problem with these two flabby, wordy sentences rests with the verbs ("will be done" and "will occur"). Though grammatically correct, these verbs are empty: they shoulder too little information load because the action they imply — and the load they should carry — in fact reside in nouns ("examination" in the first sentence, "analysis" and "interpretation" in the second sentence).

But you can transform action from implicit to explicit by transforming nouns to loaded verbs (**boldface** type) and substituting loaded verbs for empty ones:

> The trout in this stream **will be examined** for parasites during summer 1992, and initial data **analyzed** and **interpreted** then.

In the above recasting, eliminating empty verbs also makes writing more concise by prompting other restructuring.

The examples that follow further illustrate how text can be invigorated (and made more concise) by transforming empty to loaded verbs:

**Flabby and wordy**

> Packaging of the majority of bareroot nursery stock is accomplished in kraft polyethylene bags.

**Invigorated**

> Most bareroot nursery stock is packaged in kraft polyethylene bags.

**Flabby and wordy**

> Monitoring of environmental conditions in cold storage on a continuous basis should be performed to ensure the freshness of the product.

**Invigorated**

> Environmental conditions in cold storage should be continuously monitored to ensure product freshness.

**Flabby and wordy**

> Following termination of exposure to toxic substances, the patient experienced considerable improvement in his physical symptoms and a lifting of his depression.

### Invigorated

Once the patient was no longer exposed to toxic substances, his physical symptoms improved considerably and his depression lifted.

Note also in the preceding examples how the invigorated versions seem more graspable, more concrete (see "Grounded Language," p. 97, to fully understand the principles at work here and why they matter).

# VOICE

Text is livelier and more informative when written in the active voice — that is, when the subject of the sentence (or clause) is the *doer*, or agent, of the action of the verb. When text is written in the passive voice, the subject is instead the *receiver* of the action. Let's analyze an example in detail to get to the bottom of voice which, like danglers, can be hard to identify.

Suppose you wrote

> After recovering the ball from the woods, the boy
> threw it to his friend.

In this sentence, it's clear that the "boy," the subject of the sentence, is responsible for both recovering and throwing the ball. The fact that "boy" is the subject *and* the doer tells you this sentence is constructed in the active voice. Note that "ball," "it" (a referent to "ball"), and "friend" are the receivers of the action.

But you might have written instead

> After the ball was recovered from the woods, the
> boy threw it to his friend.

In this recasting, the original phrase "After recovering the ball from the woods" has been turned into the passive clause "After the ball was recovered from the woods." Note that "ball," the subject of the clause, is the receiver, not the doer (**boldface** for emphasis):

> After the **ball was recovered**...

Although grammatically correct, this sentence begins to show the loss of clarity symptomatic of writing in the passive voice. Although it still tells us who threw the ball, it doesn't tell us who recovered it. We might assume it was "the boy," but we don't actually know that.

But what if you'd written

> After the ball was recovered from the woods, it was thrown by the boy to his friend.

In this recasting, we still don't know for sure who recovered the ball. Worse yet, the sentence is now flabby and wordy with the ill-advised transformation of the active clause "the boy threw it to his friend" to the passive clause "it was thrown by the boy to his friend."

Passive voice is often the structural culprit at the root of danglers. Recall one of the examples from "Dangling Phrases" (p. 59):

> Reviewing the enclosed police report, little evidence was found to support the allegation that my client, the accused, assaulted the plaintiff.

In this sentence, "Reviewing the enclosed police report" dangles from "evidence" — which is the subject of the passive clause beginning "little evidence was found." This clause is passive because "evidence" is the receiver, not the doer, of the action ("was found"). Indeed, the doer is absent from the sentence.

Converting the sentence from passive to active voice by

inserting the doer ("I," a referent to "lawyer") adds information, which clarifies the sentence *and* eliminates the dangler:

> Reviewing the enclosed police report, I found little evidence to support the allegation that my client, the accused, assaulted the plaintiff.

Let's examine, step by step through the following example, (1) how to determine whether a sentence is in the passive or active voice, and (2) how to transform a sentence earmarked as passive into an active one:

> There are no conclusions that can be drawn from this initial data.

Begin by eliminating the empties to help reveal the bare bones:

> ~~There are~~ no conclusions ~~that~~ can be drawn from this initial data.

Now you can see that the subject of the sentence, "conclusions," is the receiver, not the doer, of the action — which identifies this sentence as passive. Inserting a doer ("We") as the subject automatically relegates "conclusions," the receiver, to its proper position as object and transforms the sentence from passive to active:

> We can draw no conclusions from this initial data.

You could even turn the noun "conclusions" into its verb form "conclude" to further shorten and invigorate the active sentence:

> We can conclude nothing from this initial data.

For the sake of clarity and conciseness, prefer the active voice. But don't use it exclusively! Passive voice has its place in language; otherwise, it would've disappeared as a grammatical element long ago. It doesn't always cause problems and sometimes is the better way to construct a clause or sentence. Prefer the passive voice when the emphasis is not on the doer of the action, indeed where the doer is secondary or irrelevant —

> This relatively rare problem will not be dealt with here.

— or as yet another way to manipulate writing dynamics by varying sentence type. For instance, you might prefer to retain the passive version "No conclusions can be drawn from this initial data" — as long as clarity is in no way compromised — if the doer is unimportant.

See if you can expose the active (underlined) and passive (**boldface**) construction(s) before I do in each of the following three numbered sentences and then decide where recasting might clarify or otherwise improve the sentence:

> 1  In this case, the writer ill-advisedly turned the original phrase into a passive clause.

Active voice exposed (subject is the doer of the action):

> In this case, the writer ill-advisedly turned the original phrase into a passive clause.

This version might be transformed to **passive** (subject is the receiver of the action) if emphasizing "the writer" is inappropriate: [7]

> In this case, the original **phrase was** ill-advisedly **turned** into a passive clause.

◊

> 2  When operational aspects of the shelterwood and clearcut methods for regenerating forest stands are compared, five important factors must be considered.

**Passive** voice exposed (subject is the receiver of the action):

> When operational **aspects** of the shelter-
> wood and clearcut methods for regenerating
> forest stands **are compared**, five important
> **factors must be considered.**

Recast versions in which some or all **passive** has been transformed to <u>active</u> (subject is the doer of the action):

> When operational **aspects** of the shelter-
> wood and clearcut methods for regenerating
> forest stands **are compared**, land <u>managers</u>
> <u>must consider</u> five important factors.

Or

> When comparing operational aspects of the
> shelterwood and clearcut methods for re-
> generating forest stands, land <u>managers</u>
> <u>must consider</u> five important factors.

Or

> When land <u>managers compare</u> operational
> aspects of the shelterwood and clearcut
> methods for regenerating forest stands, <u>they</u>
> <u>must consider</u> five important factors.

◊

3 We guessed, from the blood on the floor, that the
  burglar had been cut on the jagged glass protruding from the window that he broke sometime during the night.

<u>Active</u> voice (subject is the doer of the action) and **passive** voice (subject is the receiver of the action) exposed:

> <u>We guessed</u>, from the blood on the floor,
> that the **burglar had been cut** on the
> jagged glass protruding from the window
> that <u>he broke</u> sometime during the night.

This sentence — reflecting a healthy mix of active and passive — warrants no recasting.

# GROUNDED LANGUAGE

Text with few empties and little passive voice will be clear rather than ambiguous, specific rather than general, and concrete rather than abstract. That is, it will be grounded. Indeed, the more esoteric your subject matter, the more crucial it is that your language be grounded.

As a simple example, if you're documenting the effects of a major pest outbreak on a forest ecosystem, don't write

> **The majority of the mortality occurred** at the end of the dry season...

when you can instead write

> **Most of the trees died** at the end of the dry season....

In the first version, the highlighted language is unnecessarily inflated, or puffed up, and (once again!) more words are used than are needed to make the point. Also note the empty verb "occurred."

In the second version, the highlighted words relate directly to concrete reality, the minimum number of words is used to make the point, and all words carry their proper load (note the power verb "died"). Readers can absorb the information effortlessly and move on *because the words never get in the way of the meaning*. Recall, from "Verb Power" (p. 88), the flabby, wordy sentence:

> Following termination of exposure to toxic substances, the patient experienced considerable improvement in his physical symptoms and a lifting of his depression.

Its recasting is invigorated and clarified because the language is better grounded:

> Once the patient was no longer exposed to toxic substances, his physical symptoms improved considerably and his depression lifted.

Technical writers tend to stray from the plainspoken to the puffed because it sounds more impressive, more "specialized" or "scientific" — as if bigger words and more expansive sentences imply bigger, more expansive thinking. In fact, the deluge of puffery dilutes meaning. More is actually less:

**Puffed up**

> The immediate post-wildfire environment imposes a sudden and drastic modification on the local microclimates associated with various animal habitat structures.

**Grounded**

> Animal habitats and their microclimates are suddenly and drastically modified after wildfire.

**Puffed up**

> In order to facilitate collaborative efforts and interdivisional cooperation of the engineering

teams selected at all seven project sites, the manager of the overall project has set, as a key organizational goal, the development and implementation of an electronic intersite communications network.

**Grounded**

The seven project sites will be connected by an electronic communications network so that engineering teams at all sites can collaborate more effectively.

More is likely to be least, and the consequences for clarity potentially most disastrous, when puffery interacts with pretentious *jargon*. Not all jargon — the technical vocabulary of specialists — is pretentious. Indeed, some is self-explanatory and immensely useful. But puffery and pretentious jargon are a deadly combination, as the following two examples from a draft environmental impact statement illustrate:

Habitat capability reductions during the second year after logging activities and cover removal are expected to cause large decreases in the ability of deer to capture the increased capacities present later on.

This sentence says something about problems that deer will have in their native habitat after logging.

All of the primitive, semiprimitive motorized, and semiprimitive nonmotorized recreation opportunities would be eliminated, and the entire area would be inventoried as roaded natural opportunities.

This sentence says something about how an area would be developed for recreation.

The preceding two example sentences may be perfectly comprehensible to the "in crowd" — the specialists who routinely write and read them. But to any other set of readers:

> These communications experts who employ specialized technical vocabularies are expert practitioners of glossolalia.

That is:

> These technical writers are speaking in tongues.

Translating text like the preceding can yield only a set of best guesses as to what the writer meant because both writer and text are floating in the linguistic stratosphere.

# Ending: After Writing the Last Word

# Editing Your Own Writing

By the time you've finished composing, and especially if your project has been a long one, you'll be burned out — and sick and tired of your material. It's only natural that you want that *&#?% document off your desk!

But resist the urge to immediately send it on its way. Instead, *shelve it!* That's right — take a break from it to let it "cool," let others not intimate with it review it during that break, and then revise and thoroughly proofread it one last time. This final phase of editing your own writing, though often agonizing because you're eager to move on to new work, is crucial to ensure the ultimate goal, clarity.

## Cool-down

Once you've finished writing — that is, once you've finished the creative act of composing text — let your document "cool" for as long as you realistically can: a morning or an afternoon if your deadline is pressing; several days, perhaps over a weekend, if you can manage that; or (best!) several weeks. The longer, the better. Because you're bound to miss a lot in the heat of the creative flow, it's vital to *plan* to catch it through cool-down.

Distancing yourself physically from your creation allows you to disengage from it mentally. And by disengaging, you effectively separate the *writing* stage from the *editing* stage — that is, the stage in which you scrutinize and revise what you've composed. Naturally, you're always doing some editing while composing. But much of that will have been done piecemeal, without an adequate overview.

Like any break, cool-down refreshes you. Then, once refreshed, you can step back from what you've been working on to truly see it. You know what *should* be there, but is it really there? You know what you *meant* to say, but did you really say it? The perspective you gain from cooling down will help you spot and correct faults in the logic flow, trim unnecessary words, and invigorate flabby text.

## Outside review

While your work is cooling, why not let someone who's "fresh" — a reviewer who serves as a surrogate reader, a sample audience — take a look at it? If you can find more than one willing person, so much the better.

Because reviewers are new to your material, they can react to it without the handicap of too much familiarity. Unlike you, they don't know what words to expect next. And unlike you, they're not emotionally vested in those words. They're in a better position to see what's really there, and if you said what you meant to. Their freshness can help you pinpoint problems you may have overlooked during the heat of composing or the piecemeal editing that accompanies it.

When choosing reviewers, think again about your intended and unintended audiences, and draw your reviewers from both pools wherever possible. If you have the option, choose a set of people with complementary skills. For example, if you've written a scientific article for submission to a refereed journal, have a colleague or two read it for accuracy of content before formal peer review and an editor[8] read it for language proficiency. If you've written a user's manual, have several "potential users" try out the instructions. Even a letter to the governor deserves a review for conciseness and clarity by a friend or family member. But whatever your document, choose reviewers who'll give you sound constructive criticism, not just a pat on the back. Flattery feels good — but can send you into print looking like a fool.

Remember, though, that even the most conscientious, insightful reviewers see the world through their own inherent set of "filters." Some of what they recommend will be reflections of their biases. So once again, use your discretion. Look on their comments as food for thought, eat what you like and what you know is good for you, and leave the rest on the plate.

# Proofreading

Error can, and does, creep into every draft at every stage. So once you've finalized text after cool-down and outside review, proofread for accuracy with utmost attention!

After each round of keyboarding, scrutinize text for typographical errors, elusive errors in content (such as incorrect numbers), words or whole lines inadvertently dropped, and inconsistencies of format. For those of you working on personal computers or word processors, *never* assume that because the text was perfect last round it's perfect still. Remember also to proofread tables, charts, graphs, and illustrations, if you have any, and to cross-check these again for consistency with the text.

Some writers rely more on the ear and prefer to proof with

a partner: one reads the latest version of the text aloud while the partner simultaneously scans the prior version. Other writers rely more on the eye and prefer to compare the latest and prior versions silently by themselves. If possible, it's also a good idea to have someone unfamiliar with your document proofread it independently.

After you've proofread for accuracy with a partner or by yourself, proof for euphony — for the rhythm and melody of the words. To do this, find a quiet spot and read your text *aloud*. If it's long, even book length, read aloud in stages, as you finalize sections or chapters. Proofing for euphony also acts as a further check on accuracy: something that sounds wrong or awkward may indicate a problem that should be pinpointed and addressed, not passed on to the reader.

Whatever your approach, proof *when you're fresh*. For most people, that's in the morning. If you do it when you're tired, you'll be appalled at what you miss — which you may only discover once it's too late: once your scientific article has been published, once your user manual is on the market, once your letter to the governor has been mailed. Don't find out the hard way that nothing looms larger than an error in print.

# Improving Your Prowess as a Writer

There's a lot you can do to fortify the vertebrate creature of writing:

- **You can read.** The more you read, the more you'll absorb and discern about both technique and artistry in writing. Indeed, the more you read, the more sensitized you'll become to the fine points that may elude you at the start. Don't make the mistake of assuming that "because it's in print, it's well written." Especially nowadays, readers are deluged with the shabby, amateurish products of nearly anyone with access to a word processor or desktop publishing. Teach yourself to discriminate.

Read with attention to the vertebrate creature of writing: to its skeleton (order), to its body mass (conciseness), and to its muscle tone (vigor). Train yourself to "edit writing" as you read it. Use analysis as a learning tool. Examine, for instance, how you might recast a sentence, reorganize a paragraph, or emphasize something important.

Read broadly — not just in the areas of special interest to you. Stretch your mind. Read general nonfiction: biographies, histories, "how to's," and so on. Read magazines, from weekly news magazines to those that popularize science, computers, and the like. Read fiction: novels, short stories. Read plays and poetry — aloud — to sensitize yourself to euphony.

- **You can pick the brains of the writers around you.** Your fellow writers provide an ideal support group. Discuss problems you've encountered; ask others for their solutions. Remember that technical writing is hard work for everyone. Don't assume that others have it down cold and you're the only one who doesn't get it. If anything, assume that they, like you, are wrestling with the vertebrate creature of writing and would welcome reinforcements.

- **You can bone up through a good reference book or a writing workshop.** Browse in a well-stocked bookstore for a book or two on writing. But shy away from the "dustcatcher," the biggest, most comprehensive book on the shelf. Chances are, you won't use it — it will overwhelm you with details and jargon, and may be hard to find things in. Choose a book that invites use, and tailor what you choose to your weaknesses (you don't need to spend time reading about what you already know) and your true level of interest (you're not trying to impress anyone). And remember, it's a *reference* book: keep it within arm's reach.

Attend a writing workshop every now and then as a "refresher," but don't expect a workshop to be a miracle cure. Prefer workshops that apply directly to the kind of writing you routinely do; for example, if you spend a lot of time preparing grant proposals, find a course designed for proposal writing. Knowledge that attaches to something in the real world tends to stick in your mind; but knowledge gained largely for the sake of virtuous self-improvement is more likely to go in one ear and out the other.

**However, nothing you can do will do more to improve your prowess as a technical writer than to *keep on writing*.** But only if you write with attention, with your brain engaged — and with full appreciation of the creative nature of the task. Technical writing *is* a fine art.

Embrace the vertebrate creature of writing as friend, not foe. Why wrestle when you can tango?

# NOTES

1  A gerund is a verbal form, ending in "-ing," that acts like a noun — for example, "*Seeing* is *believing*." Several accompanying words may follow a gerund to complete a gerund phrase.

2  An infinitive is a verb preceded by "to" — for example, "I want *to go* out now," or "You need *to think* about the consequences of your action." Several accompanying words may follow an infinitive to complete an infinitive phrase.

3  The subject is a noun or noun phrase, the predicate a verb or verb phrase.

4  As in "dangling participle." A participle is a verbal form, ending in "-ing" (present participle) or (usually) "-ed" (past participle), that acts like an adjective modifying a noun or pronoun — for example, "*Daydreaming*, my sister drove right through the red light," or "He couldn't sleep, *shaken* by what he'd witnessed." Several accompanying words may follow a participle to complete a participial phrase.

⁵ The recast versions offered here are *not* the only ones possible — you may construct others that work well. In writing, there are often many grammatical solutions to a given problem.

⁶ I use "density" here as a physicist would, to refer to mass-to-volume ratio. Dense text is concise, with a high mass-to-volume ratio. It is not "impenetrable" or "hard to understand," as some other meanings of the word "dense" imply — in fact, quite the contrary.

⁷ The recast versions offered here are *not* the only ones possible — you may construct others that work well. In writing, there are often many grammatical solutions to a given problem.

⁸ The word "editor" is a broad umbrella. Here I use it to mean an "author's editor" — someone trained to scrutinize your manuscript for clarity, organization, consistency, conciseness, sentence construction, word usage, and format.

# COLOPHON

The body of this book was designed by Dennis Stovall. The cover concept was a collaboration among the author, the publishers, and the graphic artist, Marcia Barrentine. Like all books from Blue Heron Publishing, it is printed on a pH-balanced stock.

Body text is set in ITC Stone Serif, 11/14, with various subsidiary elements in ITC Stone Informal. Section titles are 44 point Tekton with Park Avenue for capitals. Tekton is used as follows: Chapter titles, 48 points; subtitles, 24 points; initial chapter text caps, 30 points; and running heads and folios, 11 points.

All type is from the Adobe digital foundry.

ITC Stone Serif and Informal were designed in 1987 by Sumner Stone, who studied calligraphy under the late Lloyd Reynolds at Reed College in Portland, Oregon, before serving as director of typography at Adobe.

Tekton was designed by David Siegel based on the handlettering of Frank Ching. Siegel is a computer graphics and design consultant in California who studied typography under Charles Bigelow at Stanford University.

Park Avenue is a script face from 1933 with a hand-drawn quality.

Stone was chosen for its generous utility, its readability, and its adaptability to the requirements of electronic publishing. It ties the ideas of the author to the final formal process of publication, while the use of Tekton and Park Avenue link the text to the highly personal act of writing and the artistry of well-crafted words and documents — whether technical or not.